沪上·生态家

2010世博会城市最佳实践区上海案例

SHANGHAI CASE FOR URBAN BEST PRACTICE AREA, EXPO 2010

主　编　韩继红
副主编　章　颖

GUIDE BOOK

中国建筑工业出版社

沪上·生态家 导览

Eco - Housing GUIDE BOOK

目录 CONTENTS

缘起 生态之家
THE ORIGIN —— THE ECOLOGICAL HOME

生态建造　Ecological Construction
本土设计　Native Design
理念解读　Idea Analysis
本案诉求　Ecological Demands
人居变迁　Habitat Changes
百年上海　Centennial Shanghai

回望 建设历程
RETROSPECT — THE CONSTRUCTION

- 合作伙伴 Partnership
- 团队掠影 The Eco-Housing Team
- 背后故事 The Story Behind
- 竣工交付 Completion and Delivery
- 关键节点 Key Nodes
- 破土动工 Groundbreaking
- 国际遴选 International Selection
- 未来视窗 Future Window
- 乐龄之家 Apartment for the Elderly
- 三代同堂 Apartment for Extended Family
- 三口之家 Apartment for Common Family
- 青年公寓 Apartment for the Youth
- 时空长廊 Space-time Gallery
- 参见氕戋 Tour Route

体验 乐活人生
EXPERIENCE — THE LOHAS

百年上海
Centennial Shanghai

百年上海，历经沧桑，昔日的小渔村，如今已发展为现代化的国际大都市。

Shanghai has been through glorious years and cloudy times throughout a century. Having experienced lots of great changes, Shanghai is evolving from a fishing village to a thriving international megalopolis.

上海位于长江入海口，临海平原，气候适宜，是发展中国家、夏热冬冷气候区、高密度大城市的典型代表。在资源匮乏、能源短缺、污染加剧的情况下，面临着可持续发展的严峻考验。

Shanghai is a coastal plain, located in the estuary of the Yangtze River with a suitable climate. It is a typical representative of developing countries, hot summer and cold winter climate region and high density megalopolis. Shanghai is facing a severe challenge of sustainable development under the circumstances of lack of resources, energy shortages and pollution threat.

缘起 生态之家

THE ORIGIN — THE ECOLOGICAL HOME

1891年的外滩
The Bund (1891)

1930年的外滩
The Bund (1930)

1983年的外滩
The Bund (1983)

21世纪的外滩
The Bund (21st century)

人居变迁
Habitat Changes

住宅,是城市生活的基本空间。上海住宅,几度变迁,承载着人们对美好生活的追求,见证着城市为改善生活质量而作出的不懈努力。

Housing is the basic space of city life. Those changes of shanghai housing reflect people's endless pursuit for happy lives, and demonstrate the city's perpetual effort for life quality improvement.

1.上海早期本土住宅——适应江南地区夏热冬冷气候特征的传统民居。

1. The Early Shanghai Folk Dwelling (The traditional housing for adaption to the particular Southern Yangtze River Region climate characteristics of the hot summer and cold winter).

上海早期本土住宅
The Early Shanghai Folk Dwelling

2.上海近代里弄建筑——服务于激增城市人口亦中亦西的居住形态。

2. The Contemporary Shanghai Dwelling (The housing pattern of culture conflicts between the eastern housing and western housing under the circumstances of the continuously and sharply increased urban population).

上海近代里弄建筑
The Contemporary Shanghai Dwelling

3.上海当代住宅改造——应对发展中城市高消耗、高污染模式下的住宅演进。

3. The Modern Shanghai Housing (The housing evolution of reducing the high consumption and high pollution during developing).

多层住宅平改坡
The roof of Multi-story Housing changed from the flat form into the slope form

老公房节能改造
Energy saving retrofit of old apartment buildings

缘起 生态之家

THE ORIGIN — THE ECOLOGICAL HOME

4. 上海生态住宅兴起——代表21世纪城市人居可持续发展的主流实践。

4. The Rise of Shanghai Ecological Housing (The popular practice represented for sustainable development of urban habitat in 21st century).

缘起 — 生态之家 / THE ORIGIN — THE ECOLOGICAL HOME

生态别墅
Ecological Villa

多层生态住宅
Multi-storied Ecological Apartment

高层生态住宅
Ecological Residential High-rise

上海生态建筑示范工程——"沪上·生态家"的母体
(位于上海市申富路568号,于2004年9月起建成开放)

"Shanghai Ecological Demonstration Building" project—the matrix of "Shanghai Eco-Housing"(It is located in No. 568 Shenfu Road, Shanghai, and officially opened since September, 2004)

本案诉求 / Ecological Demands

"沪上·生态家",作为"上海生态住宅"应对"夏热冬冷地区、高密度、大城市"地域特点的最佳实践典范,以"诠释世博主题,展现上海本土特色、成功经验和最高水平,引领未来发展"为使命,以荣获2005年建设部首届绿色建筑创新奖一等奖和2009年住房和城乡建设部首个绿色建筑三星级标识的"上海生态建筑示范楼"为母体,在建筑形态和技术应用上进行传承创新,是代表东道主上海向2010年世博会城市最佳实践区提供的居住建筑实物案例。

As a best practice of ecological building under the circumstances of "hot summer and cold winter region, high density megalopolis", referring to the "Shanghai Ecological Demonstration Building"(the first building won the first prize of 2005 MOC Green Building Innovation Award and 2009 MOHURD Green Building Operation Label), "Shanghai Eco-Housing" interprets the theme of the Expo, displays the top level in Shanghai, leads the future development, and inherits the innovation ideas on the building form and technology application. On the behalf of Shanghai, it is also the residential practical case exhibited in the Expo UBPA.

沪上·生态家实景图
Eco-Housing

理念解读 / Idea Analysis

"沪上·生态家"遵循"天和——节能减排、环境共生,地和——因地制宜、本土特色,人和——以人为本、健康舒适,乐活——健康可持续的价值导向"的定位目标,采用因地制宜的设计原则和自主创新的关键技术,集上海和国内外之大成,在方案的策划和设计中践行了"生态建造,乐活人生"的全新生态居住理念。

Integrating local design and innovative technologies, "Shanghai Eco-Housing" follows the ideas of energy conservation, emission reduction, environmental protection, human orientation and LOHAS.

缘起 生态之家 — THE ORIGIN — THE ECOLOGICAL HOME

本土设计
Native Design

缘起 — 生态之家

"沪上·生态家"建筑设计充分汲取江南民居的传统精髓,以白墙、灰砖、黄瓦为主色调,以山墙、里弄、老虎窗等为上海住宅要素符号,以"风、光、影、绿"等本土生态手法传承和演绎为重中之重,通过老虎窗和楼梯天井设计强化自然通风采光,外挑屋檐设计满足遮阳避雨功能,景观水体生态保持使得滨水而居却流水不腐,而南阳台模块绿化墙、西墙爬藤绿化、中庭风笼垂挂植物绿化、屋顶花园绿化等无处不在的生态绿化立体配置,使整个建筑物处处草木葱郁,融入绿色盎然之中,充分体现了生态家自然和谐的居住理想。

THE ORIGIN — THE ECOLOGICAL HOME

The architectural design of "Shanghai Eco-Housing" fully inherits the essence of traditional southern dwellings.
- Hue and Material: white wall, grey brick, yellow tile.
- Traditional Native Design: gable, lanes and alleys, dormer, etc.
- Ecological Elements: Wind, Light, Shadow, Green.

Ecological Technologies Application:
- Dormer provides natural ventilation.
- Stair patio provides natural lighting.
- Overhanging eaves provides sun shade and rain shade.
- The landscaping pond keeps the water fresh.
- All the ecological green fully display the idea of "Nature and Harmony".

白墙灰砖、滨水而居
White Brick and Grey Brick Near the Water

石库门意向
Shikumen Design

老虎窗
Dormer

中庭风笼
Atrium

屋顶花园
Roof Garden

生态建造
Ecological Construction

"沪上·生态家"建筑本身便是一个满足我国绿色建筑三星级最高要求的实物展品,通过各种最新科技产品的建筑一体化应用,向我们展示了打造"节能低碳、固废再生、环境宜居、智能便捷"生态家园的技术亮点。

"Shanghai Eco-Housing" is a three-star green building in China.
Advantages:
• Energy Conservation and Low Carbon.
• Solid Waste Recycling.
• Environment-Friendly.
• Intelligent and Convenient.

缘起 — 生态之家
THE ORIGIN — THE ECOLOGICAL HOME

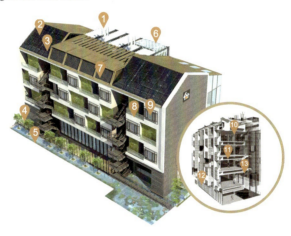

1. 屋面小型风力发电
   Wind Power Generation
2. 非晶硅薄膜太阳能发电
   Solar PV Generation
3. 平板集热太阳能热水
   Solar Heat Water
4. LED夜景照明
   LED Landscape lighting
5. 生态浮岛景观水池
   Landscape Pond
6. 中庭强化自然通风
   Natural Ventilation
7. 老虎窗天然采光
   Natural Lighting
8. 南阳台挂壁式模块绿化
   Wall-Hanging Modular Green
9. 双层窗夹遮阳帘系统
   Double Glazing Sun Shading System
10. 能量回馈节能电梯
    Energy Recovery Elevator
11. 隔热保温调温调湿墙体
    Heat Insulation Wall
12. 再生骨料绿色混凝土
    Renewable Aggregate and Green Concrete
13. 固废再生内隔墙
    Waste Reutilization Partition Wall

"沪上·生态家"技术亮点
Highlights of "Shanghai Eco-Housing"

1. 气候适应性节能围护结构——冬暖夏凉的建筑外衣

- 墙体：内墙用电厂废料脱硫石膏制成的板调节室内的温湿度，中间填充长江淤泥制成的砖，外包无机保温砂浆和隔热涂料（或隔热砂浆），给整个建筑穿上了冬暖夏凉的建筑外衣。
- 外窗：采用内外都可以打开或关闭的双层窗、中间内置可收放的遮阳卷帘，具有良好的节能效果和通风换气、采光、隔声综合性能，成为家庭窗户节能改造的样板。

1. Climate-adaptive energy saving building envelope — a building coat for warm winter and cool summer area

- Walls: Waste gypsum boards made for interior walls are used to control the indoor temperature. The Yangtze silt bricks in the middle, together with inorganic thermal insulation mortar on the façade and heat insulation coating (or insulation mortar), give the entire building a warm winter and cool summer coat.
- Exterior windows: As a model of energy saving retrofit, both sides of double-windows can be opened and closed. The sun shade curtains in the middle have not only a good energy saving effect but also a comprehensive performance including ventilation, lighting and sound insulation.

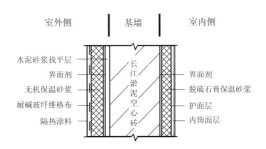

墙体保温结构
Wall Insulation Structure

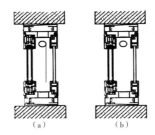

双层窗系统
Double-window System

2. 固废再生绿色建材——变废为宝的建筑材料

- 含废弃料高达60%以上的混凝土：将旧房拆除的建筑垃圾打碎后代替粗石料，用矿渣、粉煤灰等工业废料代替部分水泥生产的节能环保型绿色混凝土。
- 100%采用固废再生建材的内隔墙：利用废纸制成的保温纸芯隔墙、利用生活垃圾制成的砌块等。
- 装饰装修材料的固废再利用：采用废弃型钢焊接加工成中庭风笼钢结构，用约15万块老石库门砖砌筑建筑立面，铺砌楼梯踏面等。

2. Solid waste recycling green building materials — a technology can change waste into treature

- 60% of the high performance recycling aggregate concrete comes from waste: after old buildings were demolished, the building waste broken into pieces takes the place of rough stones. Slag and fly ash take the place of cements. Those waste materials are changed into energy-saving and environmental-friendly green concrete.
- The interior partition wall uses the recycling waste as building materials in 100%: Waste paper is used for heat insulation partition wall. Garbage is used for blocks.
- The decoration materials adopt the solid waste recycling technology: The steel cage structure of atrium is made by waste steel welding and processing technology. Walls and stairs of the Eco-Housing are made from about 150,000 old Shikumen bricks.

打碎后再利用的建筑垃圾
Recycle Building Waste

用旧砖铺砌的楼梯踏面
Old Bricks Used For Stairs

3. 非晶硅薄膜BIPV系统——更适合上海气候的太阳能发电

在建筑的屋顶和阳台上,采用了更易与建筑一体化贴合的非晶硅薄膜太阳能发电系统。与以往常见的单晶或多晶硅太阳能发电系统不同,它不需很强的阳光,只要有光线透过就能发电,特别适合太阳能资源不够丰富的上海地区。

3. Amorphous silicon film BIPV system — a more suitable solar generation system to the climate in Shanghai

The BIPV amorphous silicon film solar PV system is used on the roof and balcony. Compare to the monocrystalline silicon type and polycrystalline silicon type, it can generate the electricity even in cloudy days, which is more suitable to the places lack of solar energy, especially Shanghai.

屋顶上的太阳能发电系统
Solar PV Generation (Roof)

阳台上的太阳能发电系统
Solar PV Generation (Balcony)

4. 小型垂直轴风力发电机——屋顶上竖起来的发电大风车

这种新型的风力发电机,微风时就能启动发电,置于屋顶的钢屋架上,噪声小、寿命长、效率高,成为绿色建筑的标志性元素之一。

4. Small vertical-axis wind turbines — a large vertical windmill on the roof

The new type wind turbines set on the roof can generate electricity when the wind passes.

Advantages:
- Less Noise.
- Long Working Lifecycle.
- High Efficiency.

Those wind turbines have become one of the symbolizing elements of green building.

屋顶小型垂直轴风力发电机
Small Vertical-axis
Wind Turbines on the Roof

5. 平板集热太阳能热水系统——会隐身的太阳能热水器

跟建筑屋架完美"隐身"结合的平板集热太阳能热水器,寿命长、效率高,不仅为整楼提供50%的生活热水量,还可有效消除太阳能板带来的"光污染"问题。

5. Flat heat collector solar heat water system — an invisible solar water heater

The "invisible" flat heat collector solar water system set on the roof is perfect integrated with the entire building design, to provide the building 50% of hot water needed.

Advantages:
- Long Working Lifecycle.
- High Efficiency.
- Elimination of Light Pollution.

屋架平板集热太阳能热水系统
Solar Heat Water System

6. 能量回馈电梯——上上下下中的能发电电梯

采用了"能量回馈技术"的电梯,在上下运行过程中,多余势能和动能会转化成电能向电梯控制系统反馈,通过轿厢内置的实时监测视窗,可清晰地看到其节能30%以上的效果。

6. Energy Recovery Elevator — an elevator can generate electricity during working

The "Energy Recovery Technology" is used on all the elevators.
Principles:
- They can convert the excess potential energy and kinetic energy into electrical energy.
- The electrical energy can be fed back to the elevator control system.

Effects: Those energy recovery elevators can save the energy over 30%.

能量回馈电梯
Energy Recovery Elevator

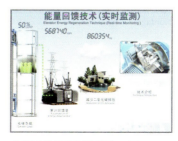

能量回馈实时监测视窗
Energy Feedback Real-Time Monitoring System

7. 生态浮床水体净化——一流水不腐的自洁景观水池

将生态浮床与地下室景观水池巧妙地结合在一起,利用植物净化收集到的屋面雨水补给景观用水,并在水景池里种植同时具有净水和景观功能的水生植物,让滨水而居却流水不腐,了却了居家的后顾之忧。

7. Water purification from the ecological floating beds — an automatic cleaning pond with fresh water

Principle:

- By using the plant purification technology, the ecological floating beds and the basement landscaping pond are integrated to purify the rainwater having been collected from the building.
- All the aquatic plants have the advantages of water purification and landscaping.

地下室景观水池水生植物
Underground Landscaping Pond

8. 墙面垂直绿化——可随时更换的绿衣服

南侧阳台墙面采用挂壁式种植模块绿化,植物在环保型纸花盆中定型培养,具有安全美观、节能生态、随调随配、成景迅速及养管简便等优势。西侧墙面为防强烈的西晒阳光,种植爬山虎等来隔热降温。

8. The vertical green wall — a diverse green clothes

The wall-hanging modular plants are decorated inside the walls of the south balcony. All the plants are living inside the environmental-friendly pots.

Advantages:

- Safe and Beautiful.
- Energy Conservation.
- Ecological.
- Easy for Change, Decoration and Maintenance.

南侧阳台墙面挂壁式模块绿化
Wall-hanging Modular Plant on the South Balcony

Lots of plants such as ivy are decorated on the west walls for heat insulation.

西侧墙爬藤绿化
Ivy on the West Walls

9. LED室内外照明——营造节能多彩居家环境

采用新一代LED光源用于室外夜景照明、室内公共区域背景照明以及局部功能性照明，兼具节能高效、色彩斑斓特性，营造环保宜居多姿氛围。

9. LED lighting — a diverse living atmosphere

LED lighting is used for outdoor night lighting, public indoor background lighting, and partial function lighting, with the characteristics of energy conservation, high efficiency and environmentally-friendly living.

Functions:
- Outdoor Night Lighting.
- Indoor Background Lighting for Public Area.
- Regional Functional Lighting.

Advantages:
- Energy Conservation.
- Environmental-friendly.
- High Efficiency.
- Harmonious.
- Gorgeous.

LED外立面夜景照明
Outdoor Night Lighting

室内节能多彩照明
Indoor Background Lighting

10. 智能监控管理系统——智能管家让生活节约又舒适

由设备管理中心、能源管理中心、环境监控中心、信息展示中心组成的智能管家，通过对能源和环境的监控，协调管理各种自动化设备的运行，在保证人体健康舒适的情况下实现能耗最小化。

10. Intelligent Monitoring Management System — a comfortable and economy life

The equipment management center, energy management center, environment monitoring center, information display center constitute the "intelligent steward".

Functions:
- Monitor the energy consumption.
- Monitor the environment.
- Run the automatic equipment.
- Secure the human health.

Advantages:
- Low energy consumption.

集成平台画面
Integration Platform

能源分析图表
Energy Analysis Chart

参观流线
Tour Route

体验 — 乐活人生 / EXPERIENCE — THE LOHAS

"沪上·生态家"位于此届世博会的创举和亮点、E片区的城市最佳实践区北部,这里由来自全球15个城市的最佳实物案例共同构成了一个集住宅、办公和休闲为一体的模拟生活街区,其中有一幢白墙青砖黄瓦、极具江南民居特色的4层楼房子,醒目地伫立在与伦敦和马德里案例比邻的居住组团的门户入口处,这就是代表东道主上海的城市实物案例"沪上·生态家"。

参观者从水陆两线都能到达浦西的城市最佳实践区,穿过南区的全球城市广场,经由中区的人行天桥向北走,就能将"沪上·生态家"的风采尽收眼底。

"Shanghai Eco-Housing", is located in the north side of Expo Zone E, UBPA. Eco-Housing, together with fifteen foreign practical cases, constitute a simulation living block. On the behalf of Shanghai, "Shanghai Eco-Housing" is a four-story high southern style building decorated with white wall, cyan brick and yellow tile.

Location:
- Near the London case pavilion and Madrid case pavilion.

Tour Route:
- UBPA, Zone E, Expo Puxi site ⟶ Global Urban Plaza ⟶ Pedestrian Bridge ⟶ Eco-Housing.

Tour Direction:
- South Part of UBPA ⟶ North Part of UBPA.

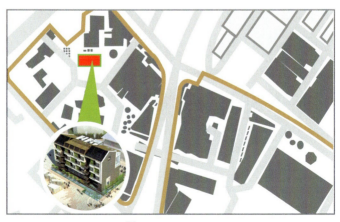

"沪上·生态家"区址图
Location of "Shanghai Eco-Housing".

"沪上·生态家"建筑面积3017m², 地上4层, 以"住宅科技时空之旅"为主题, 布置了由"过去、现在、未来"三部曲式的展示内容, 带领观众从"过去"的时间之桥和警示剧场, 聆听上海近现代以来的人居变迁; 从"现在"的"青年公寓、两代天地、三世同堂、乐龄之家"的实物居家模式展示, 看到绿色科技带给每个人的理想乐活家园; 从"未来"的多媒体畅想, 切身感受绿色低碳行为模式下城市生活的美好未来。

Building Area: 3017 square meters.
Pavilion Theme: "Ecological Building, Lifestyle of Health and Sustainability".
Exhibition Theme: "Space-Time of Residential Science and Technology".
Exhibition Contents: "Past", "Now", "Future".

- **"Past":** Through the bridge of time, warning theater, etc, this part of story is mainly about the habitat change in contemporary and modern Shanghai.
- **"Now":** Through the theme across all stages of people's life, this part of story shows the green life created by science and technology innovation.
- **"Future":** Through a video about future, this part of story brings visitors a beautiful image of green and low-carbon residential living in the future.

上海乐活人生体验
Experience of LOHAS

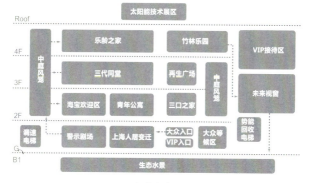

参观流线图
Tour Chart

时空长廊
Space-time Gallery

接下来,就让我们一起深入其中,细细领略"沪上·生态家"的乐活人生体验。

Now let's begin our tour in "Shanghai Eco-Housing" and experience the LOHAS.

体验 — 乐活人生 / EXPERIENCE — THE LOHAS

一层平面图
Plan of the 1st Floor

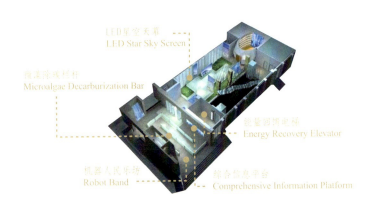

一层参观亮点
Highlights of the 1st Floor

首先映入眼帘的是丰富多彩的室外等候区。3位美女机器人穿着传统的中国式旗袍,面带微笑为您演奏中阮、扬琴、葫芦丝3种中国民族乐器。除点歌外,您还能与她们进行互动,学习器乐的演奏。在优美音乐的伴随下,您将穿过"会呼吸的围栏——微藻除碳栏杆",它能将更清新的空气带到您身边。同时,通过墙面巨大的综合信息屏,您能够实时了解生态家的能耗使用和室内环境状况。

Before entering in the Eco-Housing, "Robot Band", "Microalgae Decarburization Bar", "Large Information Screen" are decorated in the waiting hall. Visitors can have a primary understanding of the Eco-Housing in this part of exhibition.

- Robot Band - 3 women robot musicians of the band dressed in Chinese cheongsam play in three kinds of Chinese folk instruments for visitors such as a plucked stringed instrument, dulcimer and cucurbit flute. In addition, those robots can also teach visitors how to use those instruments.
- Microalgae Decarburization Bar- the bars can breathe. Those bars can purify the air and make the waiting area with fresh air.
- Large Information Screen—Visitors can learn the real-time energy consumption and indoor environment.

机器人乐队
Robot Band

微藻除碳栏杆
Microalgae Decarburization Bar

体验 乐活人生

EXPERIENCE — THE LOHAS

　　进入一楼展馆后，您首先将穿越"时间之桥"。在这个带有江南园林九曲桥元素的展区，您将通过互动触摸屏了解到上海这座城市在改善居住质量、建设人居生态环境中所取得的历史成绩。建国60年来的上海人居建设成就将为您一一道来。在这里，您还能与地面上的"追鱼"和草地间的点点"萤火虫"互动，不过您千万要小心点，因为鱼儿有可能会被您的脚步吓得到处乱窜呢。

After entering in the exhibition hall, the "Time Bridge" is shown to visitors first. In this novel gallery designed like a zigzag bridge of southern Chinese garden, visitors can learn the improvement of dwelling house and historic achievements from the videos and touch screens. Here is a story about the achievements of dwelling house construction in Shanghai since China was founded in sixty years ago. The "Time Bridge" floated above a rolling lawn shows the people' pursuit to the nature by the "Fish Chasing" interactive multimedia technology and firefly effect technology in the lawn at night. Perhaps fishes may be scared by your steps.

时间之桥
Robot Band

　　紧随而之的是"警示剧场"。在这个圆形的小剧院中，您将通过影像直面一些生活中看似微小的问题，其实已经深深影响了整个城市甚至世界的可持续发展。那么接下来的生活我们该怎么去改变这种现状呢？"沪上·生态家"的出现，正好回答了这一问题。现在带您慢慢去解开这个谜题。

In this small circular theater, visitors can learn a story that every little thing has influence to the entire world, even the sustainable development. How can we deal with this phenomenon? "Shanghai Eco-Housing" will give you all the answers.

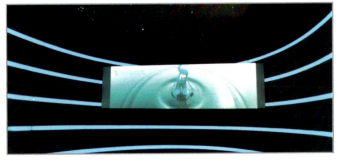

警示剧场
Warning Theater

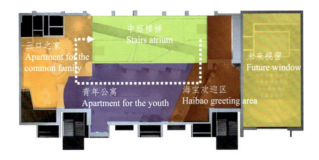

二层平面图
Plan of the 2nd Floor

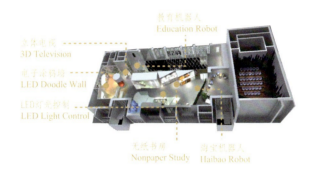

二层参观亮点
Highlights of the 2nd Floor

  从警示剧场出来，便是通向二楼的由绿萝垂挂掩映的中庭风笼，它是用废弃钢材回收处理后拼装焊接而成的，踏着由旧砖头砌筑的楼梯台阶穿过它时，您将能感觉到令人心旷神怡的丝丝凉意与徐徐清风。

  步入二层，您将看到一个完全不同的世界：若您是一位时尚而充满活力的年轻人，我们为您精心准备了专属的青年公寓；若您拥有的是幸福的三口之家，我们也为您准备了欢乐融融的三口之家，您还可以带着您的孩子与可爱活泼的海宝机器人进行互动呢。

After walking through the atrium wind cage, visitors could get to the second floor, where the Haibao Robot is waiting.

青年公寓
Apartment for the Youth

青年公寓以"高效办公、自在生活"为主题,展示无纸化阅读和LED多功能照明营造的SOHO时代.

在青年公寓,我们提倡的是一种时尚与环保相结合的生活理念。电纸书将替代传统书籍,大大减少了对森林树木的砍伐破坏。同样的,LED灯代替传统的灯具,这种灯具除了可以变换不同的颜色之外,还可以节约近2/3的电量,可谓既环保又美观时尚。这里既是您舒适的卧室,又是便捷的办公室,同时只要您愿意,这里还是最佳的举办Party的地方呢。

体验 — 乐活人生 / EXPERIENCE — THE LOHAS

As the theme of "High-Efficient Work, Comfortable Life", this part of exhibition demonstrates the SOHO mode, with the display of non-paper reading technology and LED lighting.

Idea: Fashion and Environmental-friendly.

Highlights:
- E-books instead of traditional books are provided to visitors, which plays an important role on the forestation.
- LED lighting takes the place of traditional lighting, with the advantages of energy-saving, environmental-friendly and fashionable.
- The bedroom can provide occupants different styles of life. It can be changed into an office as well as a place for holding parties.

青年公寓 / Apartment for the Youth

无纸书房 / Non-paper Study

三口之家
Apartment for Common Family

三口之家以"健康环保、时尚欢乐"为主题，展示电子涂鸦墙、3D电视、灯光环境控制系统营造的贴心的生活氛围。

在三口之家里，我们更加注重的是家庭的欢乐融洽与健康环保。您可以告别特制眼镜，在家享受3D电影盛宴。配合您的不同心情，家庭照明管家能够变化各种灯光模式，让您感受到不同风格的照明。在客厅中的电子涂鸦墙上，您的孩子可以用手指在上面随意描绘各种图案。而只需手掌在墙上轻轻擦拭，即可恢复最初的洁净。卡通造型的教育机器人，还能与您的孩子进行简单对话，寓教于乐。

As the theme of "Health and Environment Protection, Fashion and Happiness", this part of exhibition demonstrates an intimate living atmosphere by using the technologies of LED doodle wall, 3D TV, lighting environmental interactive systems.

Idea: Happiness and Harmony, Health and Environmental-Friendly.

Highlights:
- Visitors can enjoy the 3D movie without wearing special glasses.
- The domestic lighting can be changed into several different styles.
- Children can draw whatever they like on the LED doodle wall.
- The educational robot with the cartoon appearance is a good teacher for children.

3D电视机
3D Television

家庭照明管家
Domestic Lighting System

电子涂鸦墙
LED Doodle Wall

教育机器人
Educational Robot

体验 乐活人生 | EXPERIENCE — THE LOHAS

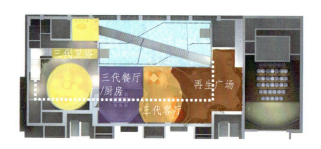

三层平面图
Plan of the 3rd Floor

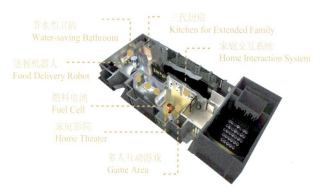

三层参观亮点
Highlights of the 3rd Floor

拾阶而上，来到三楼后，您首先看到的是一面特质的墙。它再现了"沪上·生态家"为将"城市垃圾变废为宝"所使用的各种绿色墙体材料：长江淤泥砖、混凝土空心砌块、粉煤灰砌块、回收旧砖等，充分体现了倡导节材、变废为宝的理念。

Regeneration Plaza

In this part, a special wall comes from green material is provided for visitors to experience.

Idea: material conservation, waste recycle.

Green Materials:
- Yangtze silt brick.
- Concrete hollow block.
- Ash block.
- Recycle brick.

再生广场
Regeneration Plaza

三代同堂
Apartment for Extended Family

三代同堂，则以"亲情沟通、节水环保"为主题，展示家庭信息交互系统、水处理中心、智能厨房等营造的高效能居住空间。

穿过再生广场，步入三代同堂的客厅，您可以与家人一起躺在沙发上，共同享受影音灯光完美结合后所带给您震撼的视觉盛宴。而超便捷的家庭交互系统能将您对家人的爱完美传递；多人互动游戏桌，结合了日常的电子游戏和娱乐活动，增加了家庭生活的乐趣。

As the theme of "Family Communication, Water Conservation and Environmental Protection", this part of exhibition demonstrates a high efficient living area, with the interactive screen of family information, water treatment center and intelligent kitchen.

Sitting Room

Highlights:

- The movie integrated with diverse lighting will give visitors a shocking visual banquet.
- The home interaction system can share the information not only swift but also convenient.
- The electronic game table provides a fantastic leisure life with electronic games and education.

体验 乐活人生

EXPERIENCE — THE LOHAS

家庭影院
Home Theater

家庭交互系统
Home Interaction System

多人互动游戏桌
Game Table

当您进入厨房,您将会被这个充满未来感的地方所吸引。在这里您将会感受到烹饪菜肴是一种享受,大家庭的繁重家务不再令人望而生畏!所有的抽屉、吊柜,只需您轻轻拍动,便可自由伸缩,既节省了空间,又增加了烹饪的乐趣。还有许多智能的家用设备,如智能冰箱、自动清洁的环保碗盘柜、智能化的烹饪仓等等。更加有意思的是,这里还有一个送餐机器人,它除了能将烹饪好的菜肴送到您的餐桌上,还能跟您进行语音的交流,甚至还能为您点餐呢。除此之外,这还是一个节能环保型的厨房,燃料电池机组安装于整体厨房内,它也是住宅的高效能源中心,除了能为厨房,还能为整个住宅提供热水与用电。

Intelligent Kitchen

In this part, visitors can experience a real futuristic kitchen.

Highlights:

- All the drawers, cabinets can be opened by a gentle touch.
- The intelligent refrigerator, automatic cleaning cabinet and intelligent oven make the cooking more convenient and relaxed.
- A food delivery robot can provide not only food ordering but also vocal communication.
- Fuel cell is the engine of the kitchen, which is energy-saving as well as environmental-friendly.
- The kitchen can also provide the Eco-housing hot water and electricity.

送餐机器人
Food Delivery Robot

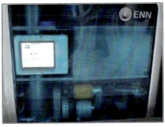

燃料电池
Fuel Cell

三代厨房
Intelligent Kitchen

离开厨房步入的便是"家庭水处理中心",在这蓝色的海洋中,向您所展示的是我们为节约水资源而作的努力。首先映入眼帘的是由多国水字组成的地面,唯美浪漫。高科技节水的未来卫生间拥有一体化坐便器,它能够自动感应冲洗,还能节约近30%的用水量。水滴状的电脑互动墙,让您了解一些有关节水的小常识。具有趣味性的家庭水处理演示中心,能将家用污水净化,这些净化后的水还能养鱼呢。

Water Treatment Center

In this part, the blue ocean environment tells visitors a story about the efforts on the water conservation.

Highlights:

- The patterns of the floor are water characters from different countries, which makes the whole area wonderful.
- The high-tech integrated toilet can provide washing by automatic induction, which can save water over 30%.
- The drop shape interactive wall provides visitors knowledge of water conservation.
- The water treatment demonstration center can purify the water.

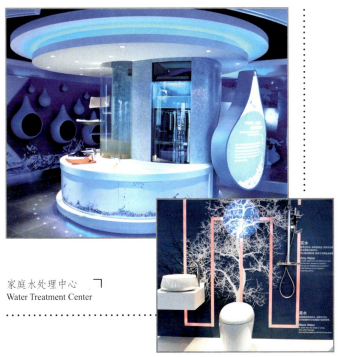

家庭水处理中心
Water Treatment Center

乐龄之家
Apartment for the Elderly

体验 | 乐活人生

EXPERIENCE — THE LOHAS

四层平面图
Plan of the 4th Floor

- 乐龄客/餐厅 Sitting Room for the elderly
- 乐龄护理 Bedroom for the elderly
- 乐龄书房 Study for the Elderly
- 竹林乐园 Bamboo Forest

- 乐龄厨房 Kitchen for the elderly
- 温馨照明 Solid Lighting
- 乐龄生活画布 Picture of the Elderly Life
- 新能源展示区 New Energy Area
- 健康监测椅 空气质量分析器 Health Monitoring Chair / Air Quality Analysis Equipment
- 护理型卫浴 Nursing Bathroom
- 护理机器人 心血管疾病监测床 Nursing Robot / Nursing Bed

四层参观亮点
Highlights of the 4th Floor

乐龄之家，以"安全、舒适、亲情"为主题，通过健康监测设备、固态照明、无障碍卫浴等，为城市里越来越多的老年人量身定做居家养生之所。

到了四层就是专为老年人悉心打造的乐龄之家。给老年人以安全、舒适与亲情是未来城市的责任与义务。在这里，不论是照明、卫浴还是床铺，只要您能想到的，我们都为老年人量身定做了特殊的设备。

在乐龄客厅，您将看到绿色节能的固态照明，还有在看电视或者休息的时候就能够完成健康检测的远程健康监测椅，还有可爱的家用空气分析器，让您能够时时知晓自己所处环境的空气状况。

As the theme of "Safety, Comfort, Family", this part of exhibition demonstrates a health maintenance place for the elderly by using healthy monitoring equipments, solid lighting, barrier-free bathroom.

Sitting Room

Highlights:
- All the equipments are specially designed for old people.
- The solid lighting in the sitting room can save energy.
- Occupants can do the health test by the remote health monitoring chair during watching television.
- The air quality analysis equipment with a lovely appearance can show the latest condition of air quality.

体验 乐活人生 EXPERIENCE — THE LOHAS

乐龄客厅
Sitting Room for the Elderly

家用空气分析器
Air Quality Analysis Equipment

乐龄厨房则是对传统厨房的极大简化，升降灶台、保温小推车、水槽便捷坐椅，凸显了高科技、人性化的设计理念；而随手可及的110、120、119安全报警系统，又为老年人的居家安全提供了实时的救援保障。

Kitchen

The kitchen for the elderly is a great simplification of the traditional kitchen.

Highlights:

- The application of lift stove, thermal preservation cart and special designed chair integrated with the sink fully demonstrates the idea of high-tech and human-orientation.
- Special buttons corresponded to security alarm systems (110,120,119) ensure old people's safety as well as provide urgent solutions.

乐龄厨房
Kitchen for the Elderly

乐龄卫浴拥有可升降的台盆、可升降的坐便器辅助起身装置，能够帮助行动不便的老人。遥控操作的升降坐席，可在浴缸中升降，让老年人也能享受沐浴的乐趣。

Bathroom

All the design in this area is for old people.

Highlights:

- The wash basin can rise and fall.
- The lifting function of the pedestal pan can help old people to stand up.
- Old people can also enjoy the shower by the lifting function.

乐龄卫浴
Bathroom for the Elderly

接下来便是乐龄卧室了。在这里，您将看到一个家用监控机器人，它的手臂可以灵活转动，除了帮老年人取东西、开门、倒水之外，还能自动遥控其他家用设备，可谓是老年人的贴身机器管家。特制的心血管疾病监测床则是针对一些常年卧病在床的老年人以及残疾人，以解决他们居家护理生活上的不方便。

Bedroom

Highlights:

- A domestic monitoring robot can help old people to take things, open the door, even control the household appliances.
- The Cardiovascular disease monitoring bed can take good care of old people and disabled people.

乐龄卧室
Bedroom for the Elderly

从卧室出来后，映入眼帘的则是由百位上海老人共同描绘的一幅壮观的乐龄生活画卷，体现了沪上老人丰富多彩、其乐融融的晚年生活和对"沪上·生态家"展现乐龄生活模式的参与和支持。

In this part, visitors can find 100 old people's daily life in a magnificent picture, which show old people's diverse life in Shanghai.

百老写真图
Picture of the Elderly Life

最后步入的竹林乐园将生态绿化、自然光引入室内，运用竹林和园林造景，凸现中国居住文化中人与自然和谐的主题。在休息互动区，采用FLASH动画、游戏、宣传片等多样化方式，全面展示太阳能、风能、生物质能、燃料电池与氢能的技术原理和应用。您可以通过玩有趣的游戏，体验有趣的新能源；专门制作的介绍未来能源的3D影片，您能够全面了解清洁能源的发展趋势、前瞻技术和广泛应用。

Bamboo Forest

Theme: Harmony between human and nature.

Highlights:

- The artificial forest landscaping design is integrated with ecological plants and natural lighting technology.
- Flash animations, games and demonstration movies fully demonstrate the technological principles of solar energy, wind energy, biomass energy, fuel cell and hydrogen energy.
- The special 3D movie tells the visitors a story about the prospects, tendency and development of future clean energy.

竹林乐园
Bamboo Forest

未来视窗 / Future Window

从四楼下至小剧场——未来视窗区,您可观看回味"沪上·生态家"建设和可回收建筑垃圾再生利用的短片,用"沪上·生态家"对世界的贡献来回应城市化带来的挑战,切身感受绿色低碳行为模式下城市生活才将有的美好未来。

Future Window

In this part, visitors can watch a short movie

4 parts of the movie:

- The construction of "Shanghai Eco-Housing".
- The process of waste recycle and reutilization.
- The achievements "Shanghai Eco-Housing" and the challenge of the urbanization.
- A much better future life created by green technologies and low-carbon action.

体验 乐活人生

EXPERIENCE — THE LOHAS

未来视窗
Future Window

至此,"沪上·生态家"整个参观过程画上了圆满的句号。
Thus, the entire show of "Shanghai Eco-Housing" has been finished.

国际遴选
International Selection

回望 建设历程
RETROSPECT — THE CONSTRUCTION

"沪上·生态家"是一个从方案、设计到建成历时近四年的项目,虽然只有3000多平方米的建筑面积,却是历经了数十次的方案修改比选并不断演进之后,于2008年3月通过了国际遴选委员会的案例评审,作为世博会城市最佳实践区代表上海城市最佳实践的实物案例,从平面图纸变身为立体工程如今屹立于世博会,与伦敦案例和马德里案例共同组成模拟街区的居住组团。

The total term of "Shanghai Eco-Housing" program including scheme phase, design phase and implementation phase lasted four years. In spite of only 3000 square meters large, Eco-Housing experienced over 10 times program changes. As the only building exhibit of Shanghai, "Shanghai Eco-Housing" passed the case international selection in the march 2008. Meanwhile, Eco-Housing, together with London case pavilion and Madrid case pavilion, constitute the residential building block.

破土动工
Groundbreaking

2008年9月10日,随着开工典礼的隆重举行,"沪上·生态家"成为世博会城市最佳实践区首个动工的案例。上海世博局领导出席并宣布了"沪上·生态家"的正式开工。

The grand opening ceremony of construction was held on September 10th, 2008. Thus, "Shanghai Eco-Housing" is the first case started construction in the UBPA. The Leaders from Expo Bureau attended and announced the start of construction.

开工典礼
Opening Ceremony of Construction

关键节点
Key Nodes

自2008年9月始的一年多时间，经历了基坑围护和开挖、基础大底板浇筑及各层的结构浇筑，"沪上·生态家"终于在2009年1月完成了结构封顶，整体形态初露端倪。

September 2008 — September 2009
- Foundation envelope
- Foundation excavation
- Baseboard pouring
- Storey structure pouring

January 2009
- Structure capping

回望 建设历程

RETROSPECT — THE CONSTRUCTION

基坑围护
Foundation Envelope

基坑开挖
Foundation Excavation

基础底板浇筑
Baseboard Pouring

一层结构浇筑
Structure Pouring (1st Floor)

二层结构浇筑
Structure Pouring (2nd Floor)

三层结构浇筑
Structure Pouring (3rd Floor)

四层结构浇筑
Structure Pouring (4th Floor)

结构封顶
Structure Capping

现场工作
Field Work

外墙饰面完成
Decoration Completion of the Building Facade

入口和LOGO
Entrance and LOGO

竣工交付
Completion and Delivery

2010年4月17日,"沪上·生态家"竣工仪式在城市最佳实践区现场举行,3天后即投入了试运行。

The completion ceremony was held in the UBPA on April 17th, 2010. And the rehearsal operation started three days later.

回望 | 建设历程

RETROSPECT — THE CONSTRUCTION

竣工仪式
Completion Ceremony

背后故事
The Story Behind

"沪上·生态家"在建设过程中充分体现了节能减排、资源回用的技术特点,我们精心选取了一些建成后无法体验与感知部位的图片,回顾建设中那些精彩的瞬间。

"Shanghai Eco-Housing" fully demonstrates the innovative technologies of energy conservation, emission reduction, and resources recycle.

Let's start to review the process of construction and see the following moments:

再生骨料混凝土的生产和使用
The production and utilization recycle aggregate concrete

生产再生骨料的原材料——旧混凝土
The Raw Material — Waste Concrete

再生骨料生产线——粉碎旧混凝土块
Production Line — Concrete Smashing

基础混凝土浇捣
Baseboard Pouring

拆模后的上部结构混凝土
Concrete on the upside

多功能PC阳台的预制和吊装
The prefabrication and hoisting of multifunctional PC balcony

多功能预制阳台模板组装
Assembly of Multifunctional Balcony

多功能预制阳台工厂制作
Production of Multifunctional Balcony

回望 建设历程

RETROSPECT — THE CONSTRUCTION

多功能预制阳台现场吊装
Hoisting of Multifunctional Balcony

多功能预制阳台安装就位
Installation of Multifunctional Balcony

太阳能和风力系统
Solar PV generation system and wind turbine generation system

太阳能屋面效果
Solar PV Panel

太阳能热水箱安装
Installation of Solar Heat Water System

屋顶风力发电机
Wind Turbine on the Roof

太阳能屋面内部
Inner Side of Solar PV Panel

团队掠影
The Eco-Housing Team

"沪上·生态家"的创意研发、展馆建设、策展布展、运营准备前后历时四年,是参与其中的核心团队和合作伙伴集体智慧的结晶,更离不开行业知名专家的把关、各级领导的关心以及各赞助单位的鼎力支持,在此铭记如下,以飨读者。

With the four-year help of the core team, all the partners, leaders, sponsors, especial all the well-known experts, the creativity development, pavilion construction, exhibition arrangement and operation preparation of "Shanghai Eco-Housing" were successful.

组织领导:
Organizer:

上海市城乡建设和交通委员会
Urban-Rural Development, Construction and Communication Committee of Shanghai Municipal Government
上海市科学技术委员会
Science and Technology Committee of Shanghai Municipal Government
上海世博会事务协调局
Shanghai World Expo Coordination Bureau

总体方案策划、工程技术支持和世博会运营管理:
Scheme Planning, Technology Support, Operation Management:

上海市建筑科学研究院(集团)有限公司
Shanghai Research Institute of Building Sciences (Group) Co.,Ltd.

工程设计施工总承包:
Engineer, Procure, Construct:

上海现代建筑设计(集团)有限公司
Shanghai Xian Dai Architectural Design (Group) Co., Ltd.

建筑设计:
Architectural design:

华东建筑设计研究院有限公司
East China Architectural Design & Research Institute Co.,Ltd.

布展策划：
Exhibit Planning

同济大学
Tongji University
上海富强建筑装饰有限公司
Shanghai Fortune Building Decoration Co., Ltd.

建筑施工：
Building Construction

上海市第四建筑有限公司
Shanghai No.4 Construction Co., Ltd.

建设监理：
Project Management

上海建科建设监理咨询有限公司
Shanghai Jianke Project Management Co.,Ltd.

回望 建设历程 / RETROSPECT — THE CONSTRUCTION

合作伙伴
Partnership

可再生能源发电
Renewable Energy Generation

新奥集团股份有限公司
ENN Group Co., Ltd.
提供的产品/技术：光伏屋顶发电并网系统、南立面光伏阳台并网系统、风能发电系统、燃料电池

太阳能热水
Solar Heat Water

江阴万龙源科技有限公司
Jiangyin Wanlongyuan Science&Technology Co,.Ltd.
提供的产品/技术：平板式太阳能集热系统

围护结构节能
Energy Conservation of Structure Envelope

常熟阿兹耐涂料有限公司
Changshu Attsu-9 Paint Co.,Ltd.
提供的产品/技术：隔热涂料

上海溢勤实业有限公司
Shanghai YiQi Industrial Co,.Ltd.
提供的产品/技术:隔热涂料、除臭涂料

江苏丰彩新型建材有限公司
Jiangsu Colourful New Building Material Industry Co.,Ltd.
提供的产品/技术:隔热涂料

圣戈班伟伯绿建建筑材料（上海）有限公司
Saint-Gobain Weber Lujian Building Materials (Shanghai) Co., Ltd.
提供的产品/技术:无机饰面砂浆

上海鑫晶山淤泥建材开发有限公司
Shanghai Xinjingshan Building Materials Development Co.,Ltd.
提供的产品/技术:淤泥空心砖

昆山长绿环保建材有限公司
Evergreen group Co., Ltd.
提供的产品/技术:无机保温砂浆

上海曹杨建筑粘合剂厂
提供的产品/技术:无机保温砂浆

上海锦莱能源科技有限公司
提供的产品/技术:相变石膏板、相变材料

圣戈班玻璃有限公司
Saint-Gobain Glass Co., Ltd.
提供的产品/技术:LOW-E玻璃

陶氏化学（中国）有限公司
Dow Chemical (China) Company Limited
提供的产品/技术:XPS挤塑聚苯板

内外遮阳系统
Sun Shading System

尚飞中国
Somfy china
提供的产品/技术:遮阳系统

上海名成智能遮阳技术有限公司
Shanghai Mingcheng Intelligent Sun-shading Technology Co., Ltd.
提供的产品/技术:遮阳系统

格伦.雷文纺织科技（苏州）有限公司
Glen Raven Asia
提供的产品/技术:遮阳系统

望瑞门遮阳系统设备（上海）有限公司
WAREMA SunShading Systems (Shanghai) Co.,Ltd.
提供的产品/技术:遮阳系统

绿色工程建材
Green Materials

上海宇山红新型建材有限公司
Shanghai Yushanhong Modern Building Materials Co.,Ltd.
提供的产品/技术:粉煤灰加气混凝土砌块

上海拉法基石膏建材有限公司
Lafarge Gypsum (Shanghai) Co., Ltd.
提供的产品/技术:脱硫石膏板隔墙

上海康居化学建材有限公司
Shanghai Kangju Chemical Building Materials Co.,Ltd.
提供的产品/技术:脱硫石膏保温砂浆

上海可耐建筑材料有限公司
提供的产品/技术:脱硫石膏保温砂浆

上海钟宏科技发展有限公司
Shanghai Zhonghong Science Technology Development Co.,Ltd.
提供的产品/技术:混凝土空心砌块

浙江天仁风管有限公司
Zhejiang Terasun Air Duct Co.,Ltd.
提供的产品/技术:复合风管

楼宇智能化平台和智能家居
Building Intelligent Platform and Smart Home

施耐德电气
施耐德电气(中国)投资有限公司
Schneider Electric (China) Investment Co.,Ltd.
提供的产品/技术:楼宇自控、智能家居

智能家居
Smart Home

飞利浦(中国)投资有限公司
Philips (China) Investment Co., Ltd.
提供的产品/技术:室内照明、家用小电器、家庭影院及护理

安玛思亚洲有限公司
AMXASIA Co.,Ltd.
提供的产品/技术:智能家居

中国科学院深圳先进技术研究院
Shenzhen Institute of Advanced Technology Chinese Academy of Sciences
提供的产品/技术:家庭医疗

深圳市松本天下科技发展有限公司
SoBen Skysafe Hi-Tech Develepment CO.,Ltd.
提供的产品/技术:数字智能控制终端及防盗报警

汉王科技股份有限公司
Hanwang Technology Co.,Ltd.
提供的产品/技术:定制电子书

节能电梯
Energy-Saving Elevator

上海三菱电梯有限公司
Shanghai Mitsubishi Elevator
提供的产品/技术:电梯

上海攀杰机械有限公司
Shanghai Panjie Machine Co., Ltd.
提供的产品/技术:无障碍升降平台

LED景观照明和室内照明节能
LED Landscaping Lighting and Energy Conservation of Indoor Lighting

上海半导体照明工程技术研究中心
Shanghai Research Center of Engineering and Technology for Solid-State Lighting
提供的产品/技术:LED景观照明、部分室内照明

机器人
Rebot

上海电气集团股份有限公司中央研究院
Shanghai Electric Group Co., Ltd. Central Academe
提供的产品/技术:机器人乐队、海宝机器人、教育机器人、送餐机器人、护理机器人、心血管疾病监测床

立体绿化
3D Green

上海植物园
Shanghai Botanical Garden
提供的产品/技术:挂壁式模块绿化

室内环境控制
Indoor Environment Control

上海领锋环境科技有限公司
Shanghai Leading AirTech Co., Ltd.
提供的产品/技术:空气净化

多功能窗
Multifunctional Windows

上海申华声学装备有限公司
Shanghai shenhua acoustics equipment Co., Ltd.
提供的产品/技术:多功能隔声窗

水处理系统
Water Treatment System

上海复泽环境科技有限公司
Shanghai Fuze Environment Technology Co., Ltd.
提供的产品/技术:中水处理系统

新型装饰装修材料和厨卫设备
Decoration Materials, Kitchen Equipments

华东理工大学华昌聚合物有限公司
Huachang Ploymer Co.,Ltd.
East China University of Science&Technology
提供的产品/技术:地坪材料

升逸豪建筑新材料（上海）有限公司
Ewall Housing and Construction materials
提供的产品/技术:纸芯隔墙

兆峰陶瓷（北京）有限公司
Siu-Fung Ceramics (Beijing) Sanitary Ware Co. Ltd.
提供的产品/技术:节水器具

东陶（中国）有限公司
Toto(China)Co.,Ltd.
提供的产品/技术:节水器具

科勒（中国）投资有限公司
Kohler China Investment Co.,Ltd.
提供的产品/技术:节水器具

美标（中国）有限公司
American Standard (China) Co.Ltd.
提供的产品/技术:节水器具

回望 建设历程

RETROSPECT — THE CONSTRUCTION

荷兰福尔波地材有限公司
Forbo flooring china
提供的产品/技术:亚麻地板

阿克苏诺贝尔太古漆油（上海）有限公司
AkzoNobel Swire Paints (Shanghai) Ltd.
提供的产品/技术:涂料

上海初美实业有限公司
提供的产品/技术:涂料

立邦涂料（中国）有限公司
Nippon Paint (China) Co.,Ltd.
提供的产品/技术:清漆、抗污内墙涂料、外墙涂料

图博节能科技（上海）有限公司
TubeEnergy-Saving Technology(Shanghai)Co.,Ltd.
提供的产品/技术:地板采暖

宁波方太厨具有限公司
Ningbo Fotile Kitchen Ware Co.,Ltd.
提供的产品/技术:厨房用橱柜

上海复甲新型材料科技有限公司
Shanghai Fujia Advanced Materials Science
& Technology Co.,Ltd.
提供的产品/技术:隔热玻璃涂料

编委会 Editorial Group

主　任： 汪维
Director

副主任： 曹嘉明
Deputy Director

编　委： 倪飞，姚栋，王佳，徐之勤，张颖，廖琳，李芳，汪添，孙桦
Editors

顾　问： 宋春华，徐正忠，陈宜明，寿子琪，黄健之，丁浩，孙建平，
Advisors 　秦云，张燕平，马兴发，朱剑豪，郑传铮，陈怡，鲁英

主　编： 韩继红
Editor-In-Chief

副主编： 章颖
Associate Editor-In-Chief

鸣谢
Credits

 上海前行广告设计有限公司
Shanghai MOTION DESIGN

 新奥集团股份有限公司
ENN Group Co., Ltd.

 飞利浦（中国）投资有限公司
Philips (China) Investment Co., Ltd.

 东陶（中国）有限公司
Toto(China)Co.,Ltd.

 科勒（中国）投资有限公司
Kohler China Investment Co.,Ltd.

 宁波方太厨具有限公司
Ningbo Fotile Kitchen Ware Co.,Ltd.

 上海电气集团股份有限公司中央研究院
Shanghai Electric Group Co., Ltd. Central Academe

施耐德电气（中国）投资有限公司
Schneider Electric (China) Investment Co.,Ltd.

 上海植物园
Shanghai Botanical Garden

 上海三菱电梯有限公司
Shanghai Mitsubishi Elevator

上海半导体照明工程技术研究中心
Shanghai Research Center of Engineering and Technology for Solid-State Lighting

图书在版编目（CIP）数据

2010世博会城市最佳实践区上海案例 沪上·生态家导览/韩继红主编. —北京：中国建筑工业出版社，2010.8
ISBN 978-7-112-12362-9

Ⅰ.①2… Ⅱ.①韩… Ⅲ.①城市建设-上海市-汉、英 Ⅳ.①TU984.251

中国版本图书馆CIP数据核字（2010）第158297号

责任编辑：韦　然　邓　卫
责任设计：李志立
责任校对：王雪竹

2010世博会城市最佳实践区上海案例
沪上·生态家导览
主　编　韩继红
副主编　章　颖
*
中国建筑工业出版社出版、发行（北京西郊百万庄）
各地新华书店、建筑书店经销
北京嘉泰利德公司制版
北京中科印刷有限公司印刷
*
开本：889×1194毫米　1/32　印张：1$\frac{1}{2}$　字数：47千字
2010年8月第一版　2010年8月第一次印刷
定价：**10.00**元
ISBN 978-7-112-12362-9
（19637）
版权所有　翻印必究
如有印装质量问题，可寄本社退换
（邮政编码100037）